INCREDIBLY DISGUSTING ENVIRONMENTS™

DEPLETED & CONTAMINATED SOIL AND YOUR FOOD SUPPLY

Carol Hand

New York

Published in 2013 by The Rosen Publishing Group, Inc.
29 East 21st Street, New York, NY 10010

Copyright © 2013 by The Rosen Publishing Group, Inc.

First Edition

All rights reserved. No part of this book may be reproduced in any form without permission in writing from the publisher, except by a reviewer.

Library of Congress Cataloging-in-Publication Data

Hand, Carol, 1945–
 Depleted & contaminated soil and your food supply/Carol Hand.
   p. cm.—(Incredibly disgusting environments)
 Includes bibliographical references and index.
 ISBN 978-1-4488-8414-8 (library binding)—ISBN 978-1-4488-8419-3 (pbk.)—
 ISBN 978-1-4488-8420-9 (6-pack)
 1. Soils and nutrition. 2. Plant nutrients. 3. Soil degradation. 4. Soil erosion.
 5. Sustainable agriculture. I. Title. II. Title: Depleted and contaminated soil and your food supply.
 S596.5.H36 2013
 631.4'5—dc23

2012023106

Manufactured in the United States of America

CPSIA Compliance Information: Batch #W13YA: For further information, contact Rosen Publishing, New York, New York, at 1-800-237-9932.

CONTENTS

INTRODUCTION 4

1 WAGING WAR ON HEALTHY SOILS 7

2 SOIL EROSION AND FOOD SUPPLY 13

3 DEPLETION, CONTAMINATION, AND FOOD SUPPLY 19

4 DEPLETED SOILS AND HUMAN HEALTH 28

5 CLEAN, RICH SOIL FOR HEALTHY FOOD 36

GLOSSARY 42
FOR MORE INFORMATION 43
FOR FURTHER READING .. 45
BIBLIOGRAPHY 46
INDEX 47

INTRODUCTION

On April 14, 1935, the sun disappeared from much of Kansas, Oklahoma, and Texas. Huge, black, rolling clouds of dust carried by 40 to 50 mile-per-hour (64 to 80 kilometer-per-hour) winds swept over the landscape and darkened the skies. The storm ripped off wheat stalks and pulled out their roots. Dust piled up like black snow drifts against buildings and fences. It clogged engines and covered the insides of houses in gritty black layers. Farmers caught outdoors were suffocated as their lungs filled with dust, and cows that later ate the dust-choked grass died as their stomachs filled with mud balls. This day—later known as "Black Sunday"—was the worst of hundreds of "black blizzards" that occurred during the 1930s.

The first major dust storm had hit the Midwest on May 11, 1934. Its winds were less strong, but a vast area—some 1,500 miles (2,414 km) long and 900 miles (1,448 km) wide—was covered with a gray-brown fog. The airborne dust cloud moved eastward, blanketing first Chicago, then Cleveland, and finally Boston and New York, with Midwestern dust. Street lights came on at midday. By themselves, those two storms ripped 650 million tons (590 million metric tons) of fertile topsoil from the farms of Kansas, Oklahoma, Texas, Colorado, and New Mexico. This was the beginning of the Dust Bowl, which devastated agriculture over 300,000 square miles (almost 777,000 square km) of the Great Plains.

But the Dust Bowl really started around 1880, when farmers began to plow up the native prairie grasses that had held the soil in place for thousands of years. As tractors, combines, and other farm machinery

replaced horses, larger and larger areas could be plowed under. Every year less native prairie remained and no plant roots were left to anchor the soil. The result was uncontrolled erosion. Then, in the early 1930s, a severe drought was combined with long stretches of temperatures over 100°F (38°C). Conditions were perfect for dust storms. By 1934, about 100 million acres (40.5 million hectares) of Midwest topsoil had been stripped away, and during the 1930s, nearly three million Midwesterners fled west toward California, away from the devastation.

After a dust storm in 1936 only the tops of farm machinery remain visible on this barn lot in South Dakota. Scenes like this were common during the Dust Bowl.

The dust storms continued until 1938, when the drought finally subsided and rains began. During this time, the Soil Conservation Service (now the Natural Resources Conservation Service) was formed and significant efforts were made, for the first time, to save the soil that feeds the nation.

It would be comforting to think that people had learned their lesson, and now, more than eighty years later, we all appreciate the importance of soil to our lives and our food supply. But since the 1930s, new and even larger dust bowls continue to form around the world. A gigantic dust bowl formed in the Soviet Virgin Lands (now Kazakhstan) during the 1960s. Today it is happening in Mongolia and northwest China, which now must import the wheat they once grew. In 2005, satellite images recorded a dust storm in the Sahel region of Africa that was larger than the entire United States. In all these areas, the pattern is the same—overgrazing, deforestation, and growing crops on unsuitable lands, followed by massive erosion and dust storms.

Soil erosion continues, and it is not the only threat to our soils. Some soils suffer from nutrient depletion. Some are compacted and unable to move sufficient water and air through their structures. Still others are contaminated by salt, acids, toxic chemicals (such as pesticides), or excess nutrients. In short, the world's soils are under attack. So before you think "Who cares about dirt?" stop and consider the importance of soil. Remember that all the food we eat depends on soil, and when soils are depleted or contaminated, the entire world food supply—and the health and livelihood of its people—is put at risk.

1 WAGING WAR ON HEALTHY SOILS

Healthy soil consists of air, water, soil particles, organic matter, and living organisms. When you squeeze a handful of clean soil, it feels moist because soil holds water. It compresses because you are squeezing out the air between the soil particles. The particles themselves contain inorganic (nonliving) nutrients. Mixed into the top layers of the soil is organic matter—the breakdown products of once-living organisms. Organic matter is broken down by the fifth component—living organisms. Many soil organisms are microbes, or microscopic organisms such as bacteria. Others are larger, such as insects and earthworms.

Soil Particles

Weathering by the constant action of wind and water eventually wears down rock to form soil particles. Weathering is aided by the actions of climate and organisms. The type of soil formed depends on the kind of rock (or "parent material") and the topography of the area (whether the land is flat or steep). Under a microscope, individual soil particles fit into three size categories. The largest, grittiest particles are sand. Silt particles are much smaller and have a floury texture. Clay particles are so tiny they cannot be seen individually, and they stick together when damp. Sand and silt form much of the structure of soil, and clay helps hold it together.

The percentages of sand, silt, and clay determine the soil's texture. In a very sandy soil, the particles do not bind together, and water drains through very quickly. A soil with mostly clay is sticky and does not drain well. The most fertile soils contain about 10-20 percent clay; the remainder is composed of about equal amounts of sand and silt. A soil with this composition—plus about 3 percent organic matter—is called loam and is the best soil for growing plants.

The Living Soil

Soil microbes break down dead organic matter (such as roots, leaves, dead animals, and manure) into simpler elements that plants can use. Some bacteria also combine small molecules into larger ones. For example, plants cannot use nitrogen gas directly from the air, so bacteria combine it into compounds that plants can take up. These nitrogen-fixing bacteria live in nodules on the roots of legumes, such as peas and beans. They capture nitrogen from the air and combine it with hydrogen to form ammonia. Other bacteria then turn ammonia or its dissolved form, ammonium, into nitrates, which plants can use.

Soil Horizons

Soils have different layers, or horizons. The thickness of each horizon varies according to the soil type, location, and amount of organic matter. Typical horizons, from top to bottom, include:

O horizon – mostly organic matter, not decomposed

A horizon – topsoil, having the most organic matter and living organisms (dark brown)

B horizon – subsoil, containing minerals leached from above, such as iron, clay, aluminum, and other soluble materials (red from iron)

C horizon – parent material, or solid rock, which also accumulates soluble compounds

This section of soil shows layers, or horizons—the dark O and A horizons (topsoil), the lighter B horizon containing leached minerals, and the bottom C horizon, containing parent material (rock).

DEPLETED & CONTAMINATED SOIL AND YOUR FOOD SUPPLY

Larger soil organisms include earthworms, spiders, mites, springtails, and insects. Like microbes, each larger organism has its own job. Earthworms tunnel through the soil, breaking up particles and forming pores to channel air and water. As they ingest the soil and feed on its microbes, they release crumbly waste products called castings, which are excellent plant fertilizer. Some of the organic matter broken down by microbes becomes humus, which is about half carbon. Fertile soil contains large amounts of humus.

Earthworms are among the most important soil organisms. A square yard of U.S. cropland contains fifty to three hundred earthworms; grasslands or woodlands contain one hundred to five hundred earthworms per square yard.

Waging War by Erosion

Erosion occurs naturally by wind and water action. Healthy soil with good texture, organic matter, and protective vegetation resists erosion. But every time people disturb the soil—for example, by plowing a field, grazing livestock, bulldozing a building site, or cutting a forest—they cause erosion. These types of land use can be done carefully, so erosion is minimized, but often they are not.

The loss of topsoil—the most fertile part of the soil—makes it harder to grow healthy food plants and support plant and animal life. Eroded soil has fewer nutrients and is less able to store water. In extreme cases, overpopulation results in stripping the land of all vegetation. This leads to desertification, or desert formation. Plus, about 60 percent of eroded soil washes into streams, rivers, and lakes. This increases flooding and contaminates the water with silt and excess nutrients.

Waging War by Contamination

We don't just erode soil. We also contaminate it by adding toxic or damaging chemicals or changing soil chemistry. This lowers the soil's ability to support life. Leaching of nutrients from the soil can make the soil more acidic or can cause buildup of toxic ions such as aluminum. Acid rain and use of nitrogen fertilizers both speed up soil acidification. Soils may be contaminated by salt buildup. They

DEPLETED & CONTAMINATED SOIL AND YOUR FOOD SUPPLY

may be damaged by pollution from mining, industries, sewage, or landfills. When toxic chemicals such as heavy metals are dissolved in water, they may leach into the groundwater and later pollute crops when the groundwater is used for irrigation.

Soil Is Not Renewable

It can take up to five hundred years to make one inch of topsoil. If an inch of soil is eroded this year, about twenty generations of your descendants will live and die before that inch of soil is replaced. Yet, a careless farmer or gardener can destroy an inch of soil in days. Although soil can be renewed, it is so easy to destroy or contaminate and so hard to replace that, for practical purposes, it is really nonrenewable. So, when people destroy soil either carelessly or on purpose—that's disgusting!

Some contaminants cannot be removed from soil. The soil must be removed and replaced. Here, workers remove soil contaminated with PCBs from the lawn of a Pittsfield, Massachusetts, home.

2 SOIL EROSION AND FOOD SUPPLY

Cornell University ecologist Dr. David Pimentel says the United States is now losing soil to erosion ten times faster than it is being replaced. Other studies suggest that actual U.S. erosion rates may be even faster than Dr. Pimentel estimates. In China and India, the rate of loss is thirty to forty times faster than the replacement rate, and around the world, soil covering an area the size of Indiana is lost every year. Worldwide, 75 billion tons (68 billion metric tons) of fertile soil are lost each year to erosion, 6.9 billion tons (6.26 billion metric tons) of it in the United States.

According to a recent United Nations forum, land degradation is one of the world's most serious environmental challenges. This group says that 40 percent of the world's land is seriously degraded and infertile because of erosion. Soil fertility has declined as a result of poor farming and grazing practices and deforestation, but these factors have larger causes—particularly population growth.

Overpopulation and Soil Erosion

Between 1980 and 2011, world population increased from 4.4 to 7 billion, and we are expected to reach 9 billion by 2050. According to the UN's World Food Programme, at least 854 million of the world's people already lack enough food to live active, healthy lives. The Earth Policy Institute estimates an even greater number—925 million. We add 4 million more hungry people every year. Experts say that, in the next 50 years, the world will have to produce more food than it has in the last 100,000 years combined, just to keep up with population growth.

But as population grows, soil erosion increases, making food production more difficult. Everyone has to eat, and often, in poor

Hungry people crowd around a food distribution site in Bedessa, Ethiopia. A massive food shortage is putting tens of thousands of Ethiopians at risk of starvation.

regions, farmland is already depleted. As the area becomes more crowded, people continue to farm the land even after it is no longer fertile, and crop production declines. They expand agriculture into marginal land where the soil is already poor. They cut down surrounding forests, clearing new land for growing crops.

But only one-tenth of the world's land surface is used as cropland. Another four-tenths is rangeland, which is too dry, steep, or infertile to grow crops. This land is used to pasture the world's 3.3 million cattle, sheep, and goats. Overgrazing by these animals is especially devastating to soil. When too many animals are grazed for too long, grassland vegetation is eaten away, soil is eroded, and once-fertile grasslands become deserts. Currently, almost half of the world's grasslands are partially degraded and 5 percent are severely degraded. The most seriously affected regions are in Africa, the Middle East, central Asia, and India—which are also the regions with the highest population growth.

Effects of Soil Erosion

Soil loss damages growing seedlings by preventing or decreasing their ability to obtain nutrients. It affects the depth to which their roots

TYPES OF SOIL EROSION

Wind erosion removes topsoil in huge dust clouds. The soil falls in other areas—often hundreds or thousands of miles away.

Water erosion occurs in three ways:
- Sheet erosion carries away a thin but uniform layer of topsoil.
- Rill erosion forms small channels (rills) shallow enough to be removed by cultivation.
- Gully erosion forms much deeper channels that pose barriers to movement of animals and machinery.

Most agricultural erosion is sheet and rill erosion. A millimeter of soil (about the thickness of a pencil tip) lost by sheet erosion results in a loss of 6.1 tons of soil per acre (13.7 metric tons per hectare) of land.

can grow. Erosion decreases the soil's ability to hold water, making it harder for growing plants to absorb water. In water erosion, the lost topsoil forms sediment that can bury seedlings or prevent them from germinating. Together, all of these factors cause lower crop yields—that is, less food is produced from every acre of land. The loss of nutrients alone costs the United States an estimated $20 billion per year.

Because of severe overgrazing, the need for animal fodder in China and India is far outpacing usable land, and the land is being further overstocked. Cattle are emaciated and unproductive, and rangelands are rapidly becoming deserts. Desertification is also expanding throughout the Sahel region of Africa, the wide band of dry land between the Sahara Desert and the rain forests of the central continent. In this region, both grazing and crop production are suffering. The country of Nigeria is a good example. Between 1950 and 2006, Nigeria's human population grew four-fold while its livestock numbers increased eleven-fold. Nigerian cropland and rangeland are now becoming desert at a rate of 867,340 acres (351,000 hectares) per year.

Soil Structure Damage

Wind or water erosion acts on the top inch or two (2 to 5 centimeters) of topsoil, but cultivation reaches deeper. When soil horizons are disturbed, the soil's structure is damaged. This makes life much harder for plants, microbes, and soil animals. Moving heavy farm equipment over soil compresses, or compacts, it. The compaction destroys pores and decreases air and water flow, especially in moist soil. Sealing and crusting cause further damage. Rain falling on bare soil washes soil surface particles downward, filling the pores. A thin seal of soil particles forms on top and then dries and hardens to form a crust. This prevents seedlings from emerging and air from entering.

Controlling Erosion

In developed countries, we already use many techniques to prevent soil erosion. We can minimize wind erosion by covering soil

with vegetation and planting trees as windbreaks. To slow water erosion, we protect soil with cover crops or mulches, so that rain infiltrates the soil, rather than running off. No-till systems are best, but if soil is cultivated, it should be done in contour furrows rather than straight rows, with strips of grass or trees planted on the contours. Retention banks can be constructed to limit runoff. Water flows into channels, and soil collecting in the channels is gathered and spread back onto the fields.

While the human population continues to grow, the food supply must grow as well, or more and more people will go hungry. But for this to happen, soil erosion must be controlled on both cropland and rangeland. Otherwise, land degradation will continue and soil fertility will be further damaged, making more land unsuitable for both crops and grazing.

The peanuts on this Georgia farm are growing in a no-till, irrigated field with a windbreak in the background. Both no-till farming and windbreaks help prevent soil erosion.

3 DEPLETION, CONTAMINATION, AND FOOD SUPPLY

Sometimes the soil—or at least part of it—remains on the fields, but becomes chemically damaged. Nutrients can leach out of the soil. Soil can also be damaged by contamination from common farming practices. Toxins can be added or useful materials can build up until they become toxic. Plants grown in contaminated soil may absorb the toxic substances—and the animal (or person) who eats that food eats the toxins, too!

It's hard to measure how much of the world's soil is damaged, but scientists have made estimates. They think about 13 percent of soil worldwide is degraded in some way, most of it by erosion. A smaller fraction has chemical damage caused by nutrient depletion or contamination. Soil is degraded the world over, but the worst damage is in Africa, Asia, and Central America.

What Plants Need

Plants need seventeen basic inorganic substances, or elements, to live and grow. If even one of these elements is missing, they

20 / DEPLETED & CONTAMINATED SOIL AND YOUR FOOD SUPPLY

die. These elements are called essential elements, or essential nutrients. Those needed in the greatest quantities are carbon, oxygen, and hydrogen. All are found in water or in carbon dioxide. They are easy to obtain, so they never limit plant growth.

The other fourteen essential elements come from soil particles. Those needed in fairly large quantities (along with carbon, oxygen, and hydrogen) are called macronutrients. Some soils have too little nitrogen, phosphorus, or potassium, so these are added as fertilizers. Animals—including humans—use plant macronutrients to build the compounds essential for life, such as proteins and carbohydrates. The other essential plant nutrients are needed in very tiny quantities. They are called micronutrients, or trace elements. Animals also need many of these substances in very tiny quantities. Like plants, we use them to carry out essential life functions. For example, we use iron to carry oxygen in the blood.

People whose loved ones died from food contamination demonstrate before USDA headquarters in Washington D.C., in 2010. Speaker Nancy Donley, of Safe Tables Our Priority (STOP), lost her son to contaminated beef.

Essential Plant Nutrients

Each group of essential nutrients in the table is listed in descending order of the amounts required. Macronutrients make up more than 99.5 percent of plant mass; micronutrients make up less than 0.5 percent.

Macronutrients	Micronutrients
Carbon	Chlorine
Oxygen	Iron
Hydrogen	Manganese
Nitrogen	Boron
Potassium	Zinc
Calcium	Copper
Magnesium	Nickel
Phosphorus	Molybdenum
Sulfur	

Soil Nutrient Depletion

Fertile soil contains all the nutrients in the right concentrations needed to grow plants. Plants are very sensitive to these levels.

They give optimal (best) yields only if the soil contains optimal amounts of all nutrients. If the soil has only half the optimal level of phosphorus, but plenty of all other nutrients, the crop yield will be only half the optimal level. Lack of phosphorus will limit the yield. Because of climate, rock type, slope, and other conditions, nutrient levels vary by location. Also, nutrients levels vary by crop. Corn requires large amounts of nitrogen, phosphorus, and potassium. Legumes require high potassium and calcium levels.

When a soil contains extremely low quantities of a certain nutrient, plants grown in the soil show specific deficiency symptoms. For example, when corn is grown in a nitrogen-poor environment, the centers of its leaves turn yellow. In a low-potassium environment, the leaf edges turn yellow. And if phosphorus is lacking, the plant is stunted and its leaves are purplish.

Soils receive nutrients by rainfall, weathering of soil and rocks, plant nitrogen fixation, and breakdown of organic matter. They lose nutrients by crop harvesting, erosion, leaching, and chemical changes of nutrients from usable to unusable forms. To maintain soil fertility, the nutrient gains must equal the losses, and to increase fertility, gains must be greater than losses.

All forms of agriculture result in nutrient loss. Farmers in developed countries such as the United States routinely test soils and apply fertilizers to maintain fertility. But in many places, especially where fertilizers are rarely or never used, soils are very depleted. For example, nutrient mining (removal of more nutrients than are replaced) occurs in sub-Saharan Africa, where soils have lost about one-fifth of their nitrogen and potassium. In the eastern Amazon region of Brazil, nutrient mining results from harvesting crops and continues even when the soil is fertilized.

Types of Soil Contamination

Thousands of chemicals can contaminate soil. But several types of contamination are common—and especially harmful. These include salts, acids, and toxic chemicals.

Erosion (shown here near Melbourne, Australia) both removes topsoil and depletes nutrients in the remaining soil. For every 2.2 pounds (1 kilogram) of bread produced, 15 pounds (7 kg) of topsoil are lost.

Salinization, or salt buildup, is a serious problem in many places. In many dry inland regions, including the central United States and Australia, soils are naturally saline. Soils in coastal areas may build up salt when ocean water enters coastal groundwater. People cause soil salinization, too—usually by irrigation. In dry areas, the salty groundwater is usually too deep to touch plant roots. But irrigation without good drainage causes soil to become waterlogged, and the groundwater level rises. When salty groundwater reaches plant roots, it reduces plant growth rates and lowers yields. In hot, dry areas, irrigation water evaporates quickly, leaving its salts behind. Fertilizers further increase the salt load.

Soil acidification is much less common than salinization. Some soils are naturally acidic and acid rain makes the problem worse. Rainwater, which is already slightly acidic, is made more acidic by pollution from smokestacks and automobile exhaust. Nitrogen fertilizers can also increase soil acidity, usually in soils with excessive rainfall or irrigation or in tropical soils. Acid soils decrease plants' ability to absorb some nutrients. They make other potentially toxic substances (such as aluminum and manganese) more available. If soil acidity is very high, plants can die in the field before harvest.

Pesticides are toxic and they can also affect soil organisms. However, if used at recommended levels, they do not appear to build up in soils. Today's pesticides are short-lived; soil microbes quickly break them down and they do not linger in the soil long enough to be absorbed by plants. However, traces of pesticides can remain on the plants themselves.

DEPLETION, CONTAMINATION, AND FOOD SUPPLY / 25

The most dangerous soil contaminants are toxic heavy metal elements, including cadmium, chromium, copper, nickel, lead, and zinc. Heavy metals usually come from specific sites, such as industrial waste or mining and smelting operations. They may also occur where wastewater or municipal sewage sludge is spread on fields. Heavy metals are already elements and cannot break down further. Instead, they are taken up by plants and concentrated as they travel up through food chains. They cause long-term health problems in animals, including humans.

Here, a crop is sprayed by airplane. Pesticides are most damaging when ingested by people eating the food. Most pesticides break down fairly rapidly and don't build up in soil.

China has one of the world's most serious soil pollution problems. China currently feeds 22 percent of the world's population on only 10 percent of the world's arable land. Its soil is overworked and it uses twice as much nitrogen fertilizer as the world average. One-tenth of its farmland has been contaminated by heavy metals and other factory wastes. Before the 2008 Summer Olympics in Beijing, many old factories were moved to new, modern sites. But the factories left behind "brownfields" where the soil is contaminated to depths of 33 feet (10 meters) and pollutant levels are one hundred times higher than acceptable levels. This land must be decontaminated before it can be used. China is now importing more and more food to meet its growing needs.

Overcoming Soil Contamination

There are three possible ways to deal with soil contamination: prevention, reclamation, and adaptation. Prevention—not contaminating the soil in the first place—is by far the preferred solution because removing contaminants is so difficult. Reclamation, or remediation, involves restoring soil health by physically removing the contaminant or neutralizing it by using bacterial or chemical action. Finally, the contaminant can be kept, and adapted plants can grown in the soil—for example, salt-adapted plants can be grown in saline soils.

China is probably the world's "poster child" for failure to prevent contamination. It has no national guidelines for soil remediation. The contaminated soil in its brownfields is now often dug up and dumped in other locations so the land can be used again because restoring the soil would take too long. Needless to say, "dig and dump" is not an ideal solution. Usually, homes or commercial buildings are built on the old sites, and the new dump sites are then too contaminated for any use, including agriculture.

In less serious cases, such as acid or salt pollution, there are better remedies. Lime can be added to neutralize acidic soils. The United States has many local sources of lime, but in the tropics, where acid soils are common, lime is often very expensive. Because of the importance of irrigation in the United States, soil salinization is a bigger problem. Salt can be leached or pumped out of irrigated soil to decrease levels enough for plant growth, but this is expensive and techniques are still being developed. In some parts of the world, salt-adapted plants (such as date palms and certain varieties of sugar beets and cotton) are grown in saline soils.

Soil Resilience

Soil has resilience—it can recover after being degraded by erosion or contamination. Badly eroded soil takes hundreds of years to recover. Less eroded soil recovers in about ten years if removed from production, planted with a cover crop, and allowed to rest. Depleted soil can be replenished quickly with organic fertilizers or humus. Acidic soil damage can be reversed by liming, but soil contaminated by salt or toxic chemicals may be unusable for many years.

4 DEPLETED SOILS AND HUMAN HEALTH

Eroded, depleted, or contaminated soils can have grim consequences. The land can be lost to agriculture entirely. It can continue to be used but produce less food. Or, it can produce food that is less nourishing or even toxic. But the amount and quality of food available to people varies greatly depending on where in the world they live.

The United States Is Overfed

People in the United States and other developed countries generally have more than enough healthy food. In the world today, about 1.6 billion people suffer health problems caused by overnutrition—that is, they eat more calories than they burn, so they become overweight or obese. Obese people have shorter life spans and are more likely to die of conditions such as heart disease and diabetes. Today, 66 percent of

DEPLETED SOILS AND HUMAN HEALTH / 29

American adults are overweight; 34 percent of these are obese (more than 20 percent over their ideal weight).

Our wealth of good food depends on industrialized agriculture, which relies on large—and unsustainable—inputs of fertilizers, pesticides, irrigation water, and fossil fuel energy. This results in huge yields, producing 80 percent of the world's food on one-fourth of its cropland. Although erosion still occurs, steps are being taken to reduce it. In the last twenty years, the United States has taken one-tenth of its

Here, students at California's Wellspring Academy work out. This special school helps teens overcome obesity through fitness and nutritional training while they continue their studies.

most erodible land out of cultivation and has begun to adopt farming methods that encourage soil conservation. During this time, erosion has decreased by 40 percent and grain yields have increased by 20 percent.

The Looming World Food Crisis

Although wealthy countries have more food than they need, developing countries face chronic, long-term food shortages. Nearly a billion people suffer from chronic hunger, or undernutrition. That is, they receive too few calories to supply their energy needs. Undernourished children suffer from mental disabilities and stunted growth and often die from infectious diseases doctors consider minor, such as measles or diarrhea. In extreme cases, undernutrition leads to starvation. Others suffer serious diseases caused by chronic malnutrition. Their diets contain too little of an essential nutrient, such as protein, vitamin A, iron, or iodine. A 2012 study says that one in four of the world's children suffer from chronic malnutrition. According to Justin Forsyth of Save the Children, "Every hour of every day, 300 children die of malnutrition, often simply because they don't have access to the basic, nutritious foods that we take for granted in rich countries."

Rapidly declining soil fertility is a big part of this problem, but other factors contribute, including climate change and rising food prices. One-fourth of the world's hungry people live in India, where climate change is causing increased temperatures, droughts, and

DISEASES OF MALNUTRITION

Symptoms of malnutrition vary according to the missing nutrient. The most common deficiencies include the following:

Deficiency	Condition	Symptoms
Protein	Kwashiorkor	Fatigue, loss of muscle mass, changes in skin and hair, severe infections, protruding belly, stunted growth
Vitamin A	Vitamin A deficiency (VAD)	Blindness, viral infections, loss of appetite, bone and skin abnormalities, childhood mortality
Iron	Iron-deficiency anemia	Fatigue, susceptibility to hemorrhage, susceptibility to infections
Iodine	Iodine deficiency	Stunted growth; mental disabilities; goiter; pregnancy problems including miscarriages, stillbirths, and congenital abnormalities

greater variation in rainfall. These factors are lowering crop yields now, and future yields will likely be even lower as droughts and floods increase erosion. Regions most affected by climate change are the same regions where malnutrition is already high.

Rapidly rising world food prices are generating food boycotts around the world. Several countries—including India, Yemen, and Mexico—have nearly had food riots. Josette Sheeran, director of the UN's World Food Programme, states, "We are facing the tightest food supplies in recent history. For the world's most vulnerable, food is simply being priced out of their reach."

Fleeing the Deserts

What happens to people when erosion runs rampant and desert replaces once-arable land? To avoid starvation, many flee to places where they can survive and grow food. This is now happening across large areas of Africa and Asia. By 2020, an estimated sixty million people will leave the Sahel region of Africa, moving to North Africa and Europe. This migration is already underway, and the refugees are desperate. In 2003, Italian authorities found a boat bound for Italy that had been

Indian citizens engage in a Right to Food Campaign demonstration in New Delhi in 2010. They are demanding guaranteed food security for poverty-stricken households, which have depleted food supplies because of severe droughts.

adrift for two weeks. The passengers had run out of food, water, and fuel. Many people had died, and the survivors were too weak to throw the corpses overboard. The boat was probably from Somalia, but survivors refused to say, for fear they would be sent back. According to Lester Brown, "Somalia is an ecological basket case, with overpopulation, overgrazing, and the resulting desertification destroying its pastoral economy."

The refugee boat was not a unique occurrence. Bodies of refugees now wash ashore daily in Italy, Turkey, and Spain. And people are not only fleeing Africa. Hundreds of Mexicans per day—seven hundred thousand per year—plus migrants from Central America are crossing the desert Southwest into the United States. Many of these desperate people are food refugees who are leaving arid lands ruined by erosion and desertification.

China probably has the world's largest and most rapidly expanding deserts. Again, the major cause is overgrazing. While the U.S. Dust Bowl displaced three million people, China's dust bowl may displace tens of millions. Gigantic dust storms—about ten per year—have become an expected part of the seasons in China. On April 5, 2001, a dust storm 1,200 miles (1,800 km) wide blanketed northwestern China and Mongolia. On April 18, that same dust storm reached the United States. It formed a cloud extending from Arizona to Canada, depositing millions of tons of China's topsoil on the American West. South Koreans now identify a "fifth season" during late winter and early spring, when China's dust storms cause school closings, airline flights

34 / DEPLETED & CONTAMINATED SOIL AND YOUR FOOD SUPPLY

cancellations, and respiratory emergencies in this neighboring country.

In China, two large central deserts are expanding and meeting to form one huge desert. In Africa, the Sahara is reaching northward, squeezing more and more people into a smaller and smaller land area. Frequent dust storms in countries such as Chad, Niger, Mauritania, and Nigeria blow westward across the Atlantic, clouding Caribbean waters and blanketing coral reefs. In the 1960s, Mauritania had about two dust storms per year;

Here, Zebus cattle in Niger search for pasture. Overgrazing causes erosion and soil loss; combined with drought, it leads to devastating dust storms. This is happening in Niger and other African countries.

now it has eighty. In Iran and Afghanistan, dust and sandstorms are burying villages and grazing areas, starving livestock, and forcing people out of their homes. Even Brazil, a relatively prosperous country, is not immune. It loses $300 million per year to desertification, which affects more than 143 million acres (58 million hectares) of land.

The Future of Soil, Food, and People

According to Lester Brown, the only way to stop the encroaching deserts around the world is to control the growth of both human and livestock populations. Earth simply has too little arable land to support the numbers we are trying to support. As we add people and livestock, we will continue to increase the numbers of hungry, malnourished, and desperate people. Turning this problem around will require action by developed countries as well as the affected people. It will require stopping erosion and desertification, replenishing and protecting soils in rangelands and croplands, replanting and conserving forests, and protecting natural ecosystems. But it will also require tackling world poverty by providing basic education for all and providing health care, including family planning education and services, to people around the world. This vast undertaking will require understanding and help from all of us.

5 CLEAN, RICH SOIL FOR HEALTHY FOOD

As a citizen of the world, you have a stake in the future of soil and food. You, your children, and your grandchildren will grow up and live in the world we are now creating. Will this world continue to contaminate and erode away its healthy soil, while condemning more and more people to malnutrition and starvation? Or will we learn to grow plenty of healthy food and make sure it is available to everyone? Many methods for sustainable farming—including formation and conservation of healthy soil—already exist. It is a matter of choosing to use them and of helping to set up conditions around the world so that other countries can use them as well.

Solutions to our soil and food crisis are both large-scale (societal) and small-scale (personal). Large-scale solutions are actions taken by large farms or corporations, national governments, or the whole world. Small-scale solutions are actions we take ourselves—things as simple as food purchases, backyard gardens, or tree planting.

CLEAN, RICH SOIL FOR HEALTHY FOOD / 37

Large-Scale Solutions

For the past thirty years, Prince Charles of Great Britain (His Royal Highness the Prince of Wales) has promoted sustainability in agriculture to ensure that we will be able to support the world's growing population. In 2004, the prince established the Accounting for Sustainability (A4S) Project to solicit global cooperation toward this goal. He feels that we must completely

This neighborhood garden is the Wangari Mathai "parken" (park/garden) in Washington, D.C., named after a Kenyan environmental activist. This once-blighted area will have individual vegetable plots, trails, and a butterfly garden.

Indoor Composting

Not everyone has a yard where they can build a compost pile. But anyone can make worm compost, even indoors in an apartment. To do this, you need a box with a lid, moist newspaper strips, and kitchen food scraps (vegetables and fruits, not meat). Put the moist newspaper strips and food scraps into the box and add some red worms ("red wigglers"). Don't use earthworms! The red worms will eat your kitchen scraps and produce black, nutritious "castings," which make perfect plant fertilizer. If you buy a commercial stacking worm bin, you can also produce "compost tea," a liquid fertilizer. Worm composting is a great classroom project.

rethink the highly technological but unsustainable agriculture upon which we now depend. This means moving away from vast monocultures created with fossil fuels and toxic chemicals, and toward natural, organic methods that build soils and maintain ecosystems.

Small-Scale Solutions

Saving soil and making agriculture sustainable involves more than changing techniques. It involves changing the attitudes of farmers, governments, and individual citizens. Everyone eats—so everyone can make changes. Here are some things you

CLEAN, RICH SOIL FOR HEALTHY FOOD / 39

Farmers' markets, such as this one in Boston, Massachusetts, are excellent sources of good organic food—and it's fresh! You can also talk to the farmers and learn their exact growing methods.

personally can do to help improve soil and ensure a healthier food supply now and in the future.

Educate yourself. Keep learning about soil and food problems here and around the world. This includes becoming aware of what you eat and where your food comes from. If you can, buy food from a local food co-op or a farmers' market, and talk to the people who grew it. The most sustainable food is locally grown and organic—that is, grown without pesticides and with only organic fertilizers. Organic food is healthier because it contains no toxins and is grown in fertile, highly organic soils.

Read labels at the grocery store and buy accordingly. Was the food imported? Again, if possible, choose locally grown, organic foods. If you eat fast food, check the company's Web site. Where was the beef for their hamburgers raised? Did they cut rain forests to graze cattle, thereby contributing to worldwide erosion? If so, perhaps you should consider changing fast food outlets.

Grow your own organic food and make your own compost. You can grow herbs and vegetables even on a deck or in an apartment, setting pots under a plant light or in a sunny window. Help start a school or neighborhood garden. Make your own compost, combining carbon ("brown") sources, such as autumn leaves, and nitrogen ("green") sources, such as grass clippings and kitchen food waste. Participate in local projects to plant trees or flower gardens, adding vegetation to beautify the neighborhood and prevent erosion. Join conservation or world food organizations, such as Oxfam International, Rainforest Action Network, Soil and Water Conservation Society, Student Conservation Association, or the Nature Conservancy.

Whatever you do, remember that your individual actions—especially if you work with family, friends, and neighbors—can make a difference. Many people making small decisions every day can combine to exert a profound force for change. In this case, they can lead to healthier soil and food for you and others. They may even help combat world hunger and malnutrition.

10 GREAT QUESTIONS TO ASK AN ENVIRONMENTALIST

1. Why is soil important?
2. How can we speed up soil production in eroded areas?
3. What is the best way to restore contaminated soil?
4. How does agriculture cause water pollution?
5. How does food growing affect natural ecosystems and biodiversity?
6. How can we incorporate organic farming methods when growing food?
7. What food buying choices can we make to help keep soil healthy?
8. Does the world grow enough food to feed today's human population?
9. How large a human population can we reasonably feed without further destroying our soils?
10. What are the best ways individuals can help protect our soils and food supply?

MYTHS & FACTS

Myth: Soil is sterile.
Fact: Soil is filled with organisms, especially microbes. A cup of soil contains billions of bacteria—as many as there are people on Earth.

Myth: All soil is brown.
Fact: Soil comes in many colors—red, yellow, and orange, among others. A soil's color results from the metals it contains (for example, red soil contains iron).

Myth: Rain forest soil makes good farmland.
Fact: Rain forest soil is only a few inches deep, making it less than ideal for farming. Once the trees are cut, the soil erodes rapidly and the productive bacteria and fungi die when exposed to sunlight.

GLOSSARY

compost Organic matter decomposed by microbes and recycled as plant fertilizer.

erosion The loss of topsoil by the action of wind and water when plant cover is removed from the land.

fertility The ability of a soil to supply plants with essential nutrients.

humus A dark-colored substance in soil formed by the breakdown of organic matter; important in soil structure, fertility, and erosion control.

leaching Washing of dissolved nutrients out of the soil as water runs through it.

loam The most fertile soil type; composed of about 10 to 20 percent clay, equal parts of sand and silt, and at least 3 perent organic matter.

malnutrition The consuming of a diet deficient in protein or other essential nutrients, resulting in deficiency diseases.

overnutrition Consumption of too many calories, leading to conditions of being overweight and obese.

salinization Buildup of salt in soil, either by covering with salt water (as in coastal areas) or by repeated cycles of irrigation and evaporation.

undernutrition Hunger, or the lack of enough calories to meet the body's energy needs.

weathering The constant action of wind and water breaking down rock to form soil particles.

FOR MORE INFORMATION

Natural Resources Conservation Service, U.S. Department of Agriculture

1400 Independence Avenue SW

Washington, DC 20250

(202) 720-2791

Web site: http://soils.usda.gov/education

The Natural Resources Conservation Service (once the Soil Conservation Service) is a part of the U.S. Department of Agriculture. It carries out research, undertakes state soil surveys, and provides information and educational materials on natural resources, including soils.

Rodale Institute

611 Siegfriedale Road

Kutztown, PA 19530

Web site: http://www.rodaleinstitute.org

The Rodale Institute has pioneered organic gardening and farming for more than sixty years. It operates a large organic research farm, sharing their methods with farmers, gardeners, scientists, and policy makers around the world. Its Web site has information on organic farming, crops and soil, food and nutrition, and more.

Soil Conservation Council of Canada

Box 998

Indian Head, SK S0G 2K0

Canada

(306) 972-7293

Web site: http://www.soilcc.ca

This nongovernmental organization is Canada's major soil conservation agency. A grassroots agency working with governments, industries, and individuals, it seeks to maintain and improve soil health in Canada and to increase public understanding of the importance of soil.

World Food Programme, United Nations

Via C.G. Viola 68, Parco dei Medici

00148 Rome, Italy

Web site: http://www.wfp.org/hunger

The UN's World Food Programme is "the world's largest humanitarian agency fighting hunger worldwide." It provides food assistance during emergencies and tries to reduce chronic hunger and malnutrition by helping countries build programs to increase food security.

Web Sites

Due to the changing nature of Internet links, Rosen Publishing has developed an online list of Web sites related to the subject of this book. This site is updated regularly. Please use this link to access the list:

http://www.rosenlinks.com/IDE/Soil

FOR FURTHER READING

Aloian, Molly. *Different Kinds of Soil* (Everybody Digs Soil). New York, NY: Crabtree Publishing Company, 2010.

Gardner, Robert. *Soil: Green Science Projects for a Sustainable Planet* (Team Green Science Projects). Berkeley Heights, NJ: Enslow Publishing Inc., 2011.

Gritzner, Charles F. *Feeding a Hungry World* (Global Connections). New York, NY: Chelsea House Publications, 2009.

Hyde, Natalie. *Micro Life in Soil* (Everybody Digs Soil). New York, NY: Crabtree Publishing Company, 2010.

Hyde, Natalie. *Soil Erosion and How to Prevent It* (Everybody Digs Soil). New York, NY: Crabtree Publishing Company, 2010.

Janke, Katelan. *Survival in the Storm: The Dust Bowl Diary of Grace Edwards, Dalhart, Texas, 1935* (Dear America Series). New York, NY: Scholastic, Inc. 2002.

Marrin, Albert. *Years of Dust.* New York, NY: Dutton Juvenile, 2009.

Montgomery, Heather. *How Is Soil Made?* (Everybody Digs Soil). New York, NY: Crabtree Publishing Company, 2010.

Peterson, Christine. *Super Soils* (Checkerboard Science Library: Rock On!). Minneapolis, MN: ABDO Publishing Company, 2010.

Pollan, Michael. *The Omnivore's Dilemma for Kids: The Secrets Behind What You Eat.* New York, NY: Dial, Adopted Edition, 2009.

Sandler, Martin W. *The Dust Bowl Through the Lens: How Photography Revealed and Helped Remedy a National Disaster.* London, England: Walker Childrens (Walker Books, Ltd.), 2009.

Silverman, Buffy. *Composting: Decomposition* (Do It Yourself). Mankato, MN: Heinemann-Raintree, 2008.

BIBLIOGRAPHY

Brown, Lester. *Plan B 3.0: Mobilizing to Save Civilization.* New York, NY: W. W. Norton & Company, 2008.

Kohnke, Helmut, and D. P. Franzmeier. *Soil Science Simplified.* 4th Ed. Long Grove, IL: Waveland Press, Inc., 1994.

Melville, Kate. "Increasing Soil Erosion Threatens World's Food Supply." Science A GoGo from Cornell University press release. March 23, 2006. Retrieved November 17, 2011 (http://www.scienceagogo.com/news/20060222202002data_trunc_sys.shtml).

Reece, Jane B., Lisa A. Urry, Michael L. Cain, Steven A. Wasserman, Peter Minorsky, and Robert B. Jackson. *Campbell Biology, Ninth Edition.* San Francisco, CA: Pearson Benjamin Cummings, 2011.

Sample, Ian. "Global Food Crisis Looms as ClimateChange and Population Growth Strip Fertile Land." *Guardian*, August 30, 2007. Retrieved January 20, 2012 (http://www.guardian.co.uk/environment/2007/aug/31/climatechange.food/print).

Watts, Jonathan. "China's Soil Deterioration May Become Growing Food Crisis, Adviser Claims." *Guardian*, February 23, 2010. Retrieved November 17, 2011. (http://www.guardian.co.uk/environment/2010/feb/23/china-soil-deterioration-food-supply).

Wild, Alan. *Soils, Land and Food.* New York, NY: Cambridge University Press, 2003.

INDEX

A
acidification, soil, 24, 27

C
composting, 38, 40
contamination of soil, 6, 11–12, 19, 23–27

D
deforestation, 6, 13, 15
desertification, 11, 16–17, 32–35
Dust Bowl, 4–6, 33

E
erosion of soil, 6, 11, 13, 14–15, 16, 27
   controlling, 17–18, 29–30
   effects of, 15–17

F
farming
   and damage to soil, 4–5, 6, 11, 12, 13, 15, 19, 22–23
   sustainable, 36, 37–38, 39
food shortages, 30–31, 32–33

G
garden, starting your own, 40

H
human health and depleted soils, 28–35

M
malnutrition, 30–31
microbes in soil, 7, 8, 10

N
nutrient depletion, 6, 11, 19–20, 21–23

O
obesity, 28–29
organic food, 39–40
overgrazing, 6, 11, 15, 16–17, 33, 35
overpopulation, 11, 14–15, 18, 35

P
plants, nutrients needed by, 19–20, 21, 22

S
salinization, 24, 27
soil
   damage to structure, 17
   horizons of, 9, 17
   living organisms in, 7, 10
   resilience of, 27
   solutions for healthy, 36–40
soil particles, 7–8

T
topsoil, 9

About the Author

Carol Hand has a Ph.D. in zoology with a specialty in marine ecology. She writes nonfiction science books for middle and high school and especially likes to write about ecology and environmental science. She has also written science curricula and standardized assessments and has taught college biology.

Photo Credits

Cover © iStockphoto.com/Michaela Fehlker; p. 5 PhotoQuest/Archive Photos/Getty Images; p. 9 USDA-NRCS; p. 10 iStockphoto.com/Thinkstock/Getty Images; p. 12 © AP Images; pp. 14–15 Press Association via Associated Press; p. 18 Inga Spence/Visuals Unlimited/Getty Images; p. 20 Ryan Kelly/CQ-Roll Call Group/Getty Images; p. 23 Eco Images/Universal Images Group/Getty Images; p. 25 Ag Aviation/Flickr/Getty Images; p. 29 Justin Sullivan/Getty Images; p. 32 Raveendran/AFP/Getty Images; p. 34 Issouf Sanogo/AFP/Getty Images; p. 37 The Washington Post/Getty Images; p. 39 Education Images/Universal Images Group/Getty Images; graphics and textures: © iStockphoto.com/stockcam (cover, back cover, interior splatters), © iStockphoto.com/Anna Chelnokova (back cover, interior splashes), © iStockphoto.com/Dusko Jovic (back cover, p. 41 background), © iStockphoto.com/Hadel Productions (pp. 4, 9, 16, 21, 31, 38 text borders), © iStockphoto.com/traveler1116 (caption background texture).

Designer: Brian Garvey; Editor: Bethany Bryan;
Photo Researcher: Amy Feinberg